How to Purify Your Drinking Water

Understanding the importance of purifying water and how purified water can keep you healthy and prevent unwanted illnesses and diseases from occurring

By Stacey Chillemi

LULU EDITION

PUBLISHED BY:
Stacey Chillemi on Lulu

How to Purify Your Drinking Water

Copyright © 2012 by Stacey Chillemi

Lulu Edition License Notes

This ebook is licensed for your personal enjoyment only. This ebook may not be re-sold or given away to other people. If you would like to share this book with another person, please purchase an additional copy for each person you share it with. If you're reading this book and did not purchase it, or it was not purchased for your use only, then you should return to Lulu.com and purchase your own copy. Thank you for respecting the author's work.

Contents

How to Purify Your Drinking Water 1
Introduction 4
Chapter 1: What is Water Purification? 8
Chapter 2: Why Is Water Purification Important? 11
Chapter 3: Why Should You Purify Your Water? 14
Chapter 4: Understanding the Seriousness of Water Contamination and How it Affects Us 16
Chapter 5: Where Can You Find Information About Your Local Water System? 22
Chapter 6: What You Can Do To Protect Yourself and Your Family 24
Chapter 7: How to Purify Your Water 29
Chapter 8: How Can You Benefit from Using a Whole House Water Filter? 33
Chapter 9: How Safe Is the Drinking Water If You Get Your Water Supply from a Household Well? 36
Chapter 10: Learn How You Can Protect Your Water by Becoming an Advocate 42
Glossary 45
Resources 52
    About the Author 56
References 60

Introduction

Do you know how safe your drinking water is? Do you know what is being done to improve the security of your public water systems? Where does your drinking water come from, and how is it treated? If you get your water from a private well, do you have any idea if private wells receive the same protection as public water systems?

This informative guide will provide you with all the answers to the questions above and this ebook will teach you how you can purify your water, so you can be healthy and avoid any unnecessary illnesses or diseases.

What can you do to help protect your drinking water?

Do you know what is in your tap water? If you don't then you don't know what you're drinking. From different studies, it has been shown from some of the water samples that contaminants were found. This is not a good sign because people should be able to drink water without having to deal with impurities and related elements. This is why home water purification is so important!

The United States enjoys one of the best supplies of drinking water in the world. Nevertheless, many of us who once gave little or no thought to the water that comes from our taps are now asking the question: **"Is my water safe to drink?"** While tap water that meets federal and state standards is generally safe to drink, threats to drinking water are increasing. Short-term disease outbreaks and water restrictions during droughts have demonstrated that we can no longer take our drinking water for granted.

You would not believe some of the things found in tap water, such as pesticides and other chemicals. They were at such high levels that it created health issues for people that drank it. In other cases, some of the test results were flawed in order to get over on the government. This kept many people in jeopardy of drinking clean water.

So many people become ill from the contaminants that are found in tap water and they do not even know why they are ill. Drinking water that has contaminants is not only bad for adults; it is also harmful to children.

Lead is one of the main contaminants that are found in water systems across the country. In fact, lead is one of the key factors that cause children to develop learning disabilities.

People that consume chlorinated water are at a greater risk. Chlorinated water has been proven a serious cancer risk, like cigarettes. Sometimes there is more chlorine in tap water than you would find in a swimming pool that uses chlorine.

In addition to that, there are so many toxic substances that people do not really know about that are harmful not only to them, but also to the environment. Why should people continue to drink contaminated water only to take a chance getting sick?

If you are not purifying your water, it is probably high time that you start. This informative guide will supply you with all the necessary information to understanding the importance of purifying your water and the step-by-step techniques on how to do it, so you can protect you and your family.

HEALTHY LIVING

The Most Important Secrets You Must Learn In Order To Stay Healthy

Chapter 1: What is Water Purification?

Water purification is the process of removing undesirable chemicals, biological contaminants, suspended solids and gases from contaminated water. The goal is to produce water fit for a specific purpose. Most water is purified for human consumption (drinking water), but water purification may also be designed for a variety of other purposes, including meeting the requirements of medical, pharmacological, chemical and industrial applications. In general the methods used include physical processes such as filtration, sedimentation, and distillation, biological processes such as slow sand filters or biologically active carbon, chemical processes such as flocculation and chlorination and the use of electromagnetic radiation such as ultraviolet light.

The purification process of water may reduce the concentration of particulate matter including suspended particles, parasites, bacteria, algae, viruses, fungi; and a range of dissolved and particulate material derived from the surfaces that water might have made contact with after falling as rain.

The standards for drinking water quality are typically set by governments or by international standards. These standards will typically set minimum and maximum concentrations of contaminants for the use that is to be made of the water.

It is not possible to tell whether water is of an appropriate quality by visual examination. Simple procedures such as boiling or the use of a household activated carbon filter are not sufficient for treating all the possible contaminants that may be present in water from an unknown source. Even natural spring water – considered safe for all practical purposes in the 19th century – must now be tested before determining what kind of treatment, if any, is needed. Chemical and microbiological analysis, while expensive, are the only way to obtain the information necessary for deciding on the appropriate method of purification.

According to a World Health Organization (WHO) report, 1.1 billion people lack access to an improved drinking water supply, 88 percent of the 4 billion annual cases of diarrheal disease come from unsafe water and inadequate sanitation and hygiene, and 1.8 million people die from diarrheal diseases each year. The WHO estimates that 94 percent of these diarrheal cases are preventable through modifications to the environment, including access to safe water. Simple techniques for treating water at home, such as chlorination, filters, and solar disinfection, and storing it in safe containers could save a huge number of lives each year. Reducing deaths from waterborne diseases is a major public health goal in developing countries.

HEALTHY LIVING

The Most Important Secrets
You Must Learn
In Order To Stay Healthy

Chapter 2: Why Is Water Purification Important?

Water purification is extremely important because chlorine is intentionally added to our public water. It is not a mistake. Chlorine is known to be very effective in disinfection. Due to its toxic characteristics, it kills bacteria, microorganisms that spread water-borne infections. Nevertheless, what is toxic to bacteria is to some extend toxic to the human body especially when the body is exposed to the chlorine for a long period.

Chlorine is absorbed through the skin and thus it can easily influence our body. Studies have shown that chlorine and especially its byproducts can become a leading cause of different cancers. Such cancers include:

- Cancer of the liver
- Bladder
- Stomach
- Colon
- And many more

Moreover, under the influence of chlorine many heart diseases are developed along with anemia, high blood pressure and allergies.

Chlorine destroys protein that is vital for the healthy skin. Having the protein destroyed the skin loses its softness and becomes less supple. Chlorine can react with other elements and form different byproducts. Chloramines often change the taste of water and spoil its odor.

The authorities are obliged to use chlorine for disinfection in the public water purification systems. So each person has to decide individually whether to remove the chlorine with the help of **water filters** or to put up with it and keep one's health in danger.

When water filters are installed on the taps only part of the problem is solved. While a person is taking a shower, the pores of the skin are wide open. Chlorine is easily absorbed through the open pores. However, in the hot shower, it is also inhaled and the lungs are affected. Then chlorine transfers to the blood stream and influences all the body. The cells become altered and mutate, which is a reason for aging, tumor and cancer.

HEALTHY LIVING

The Most Important Secrets You Must Learn In Order To Stay Healthy

Chapter 3: Why Should You Purify Your Water?

Did You Know…?

Some people may be more vulnerable to contaminants in drinking water than the general population. People undergoing chemotherapy or living with HIV/AIDS, transplant patients, children and infants, the frail elderly, and pregnant women and their fetuses can be particularly at risk for infections. If you have special health care needs, consider taking additional precautions with your drinking water.

Although most drinking water in the United States is considered safe, there is increasing concern about the quality of drinking water as more and more pollutants are found in groundwater supplies. Worry about the possible health problems resulting from these contaminants is causing people to wonder what they can do to ensure the quality of their own water supply. The most strongly recommended and best solutions to the problem of a contaminated water source are either ending the practices causing degradation of the water source, or changing water sources. These options may not always be practical or may take months or years for completion. In the mean time, other solutions may be necessary if you want to stay healthy and avoid any unwanted illnesses or diseases.

HEALTHY LIVING

**The Most Important Secrets
You Must Learn
In Order To Stay Healthy**

**Chapter 4: Understanding the Seriousness of Water
Contamination and How it Affects Us**

Chemicals, heavy metals like lead and mercury, and living organisms such as bacteria and viruses can all be threats to a safe water supply.

Unintentional contamination of water because of chemical leaks or spills, natural disasters, and other causes has been a much bigger problem than deliberate contamination.

Below are the following things below contaminating your water:

- Microorganisms (wildlife and soils)
- Radionuclides (underlying rock)
- Nitrates and nitrites (nitrogen compounds in the soil)
- Heavy metals (underground rocks containing arsenic, cadmium, chromium, lead, and selenium), fluoride
- Bacteria and nitrates
- Human and animal wastes
- Septic tanks
- Large farms
- Heavy metals
- Mining construction
- Older fruit orchards
- Fertilizers and pesticides (used by you and others (anywhere crops or lawns are maintained)
- Industrial products and wastes
- Local factories
- Industrial plants
- Gas stations
- Dry cleaners
- Leaking underground storage tanks
- Landfills
- Waste dumps
- Household wastes

- Cleaning solvents
- Used motor oil
- Paint, paint thinner
- Lead and copper (household plumbing materials)
- Water treatment chemicals (wastewater treatment plants)

The Common Types of Contamination That Destroy Our Water

Microbial Contamination:

The potential for health problems from microbial contaminated drinking water is demonstrated by localized outbreaks of waterborne disease. Many of these outbreaks have been linked to contamination by bacteria or viruses, probably from human or animal wastes. For example, in 1999 and 2000, there were 39 reported disease outbreaks associated with drinking water, some of which were linked to public drinking water supplies.

Certain **pathogens** (disease-causing **microorganisms**), such as *Cryptosporidium*, may occasionally pass through water filtration and disinfection processes in numbers high enough to cause health problems, particularly in vulnerable members of the population. *Cryptosporidium* causes the gastrointestinal disease, cryptosporidiosis, and can cause serious, sometimes fatal, symptoms, especially among sensitive members of the population. (See box on Sensitive Subpopulations on page 1.) A serious outbreak of cryptosporidiosis occurred in 1993 in Milwaukee, Wisconsin, causing more than 400,000 persons to be infected with the disease, and resulting in at least 50 deaths. This was the largest recorded outbreak of waterborne disease in United States history.

Chemical Contamination from Fertilizers:

Nitrate, a chemical most commonly used as a fertilizer, poses an immediate threat to infants when it is found in drinking water at levels above the national standard. Nitrates are converted to nitrites in the intestines. Once absorbed into the bloodstream, nitrites prevent hemoglobin from transporting oxygen. (Older children have an enzyme that restores hemoglobin.) Excessive levels can cause "blue baby syndrome," which can be fatal without immediate medical attention. Infants most at risk for blue baby syndrome are those who are already sick, and while they are sick, consume food that is high in nitrates or drink water or formula mixed with water that is high in nitrates. Avoid using water with high nitrate levels for drinking.

This is especially important for infants and young children, nursing mothers, pregnant women and certain elderly people.

Lead Contamination:

Lead, a metal found in natural deposits, is commonly used in household plumbing materials and water service lines. The greatest exposure to lead is swallowing lead paint chips or breathing in lead dust. Lead in drinking water can also cause a variety of adverse health effects. In babies and children, exposure to lead in drinking water above the **action level** of lead (0.015 milligram per liter) can result in delays in physical and mental development, along with slight deficits in attention span and learning abilities. Adults who drink this water over many years could develop kidney problems or high blood pressure. Lead is rarely found in source water, but enters tap water through corrosion of plumbing materials. Very old and poorly maintained homes may be more likely to have lead pipes, joints, and solder. However, new homes are also at risk: pipes legally considered **"lead-free"** may contain up to eight percent lead. These pipes can leach significant amounts of lead in the water for the first several months after their installation.

HEALTHY LIVING
The Most Important Secrets You Must Learn In Order To Stay Healthy

Chapter 5: Where Can You Find Information About Your Local Water System?

Since 1999, water suppliers have been required to provide annual Consumer Confidence Reports to their customers. These reports are due by July 1 each year, and contain information on contaminants found in the drinking water, possible health effects, and the water's source. Some Consumer Confidence Reports are available at *www.epa.gov/safewater/dwinfo.htm*.

Water suppliers must promptly inform you if your water has become contaminated by something that can cause immediate illness. Water suppliers have 24 hours to inform their customers of **violations** of EPA standards "that have the potential to have serious adverse effects on human health as a result of short-term exposure." If such a violation occurs, the water system will announce it through the media, and must provide information about the potential adverse effects on human health, steps the system is taking to correct the violation, and the need to use alternative water supplies (such as boiled or bottled water) until the problem is corrected. Systems will inform customers about violations of less immediate concern in the first water bill sent after the violation, in a Consumer Confidence Report, or by mail within a year. In 1998, states began compiling information on individual systems, so you can evaluate the overall quality of drinking water in your state. Additionally, EPA must compile and summarize the state reports into an annual report on the condition of the nation's drinking water. To view the most recent annual report, go to *www.epa.gov/safewater/annual*.

HEALTHY LIVING

The Most Important Secrets
You Must Learn
In Order To Stay Healthy

Chapter 6: What You Can Do To Protect Yourself and Your Family

With the exception of a known accident (such as a chemical spill into the water supply), you probably would not know that you had consumed contaminated water unless you developed symptoms. To reduce your risk of consuming contaminated food or water and to be better prepared for public health emergencies affecting the water supply follow our **Helpful Emergencies Tips** listed below.

- **Do not drink water or any other beverage that looks or smells suspicious:** In general, it is not a good idea drink something when you do not know who has prepared or provided it or where it has come from.

- **Avoid Suspicious Beverage Items:** When shopping, avoid beverage items that look like they may have been tampered with-for instance, if the seal is broken or you think that the container or packaging has been opened.

- **Remember that most cases of food poisoning, including botulism, happen by accident:** Follow guidelines for preparing and cooking food safely, keeping your kitchen clean, and washing your hands and utensils. If you preserve and can foods at home, learn and follow proper canning and freezing techniques to ensure safety. Discard cans or jars with bulging lids or leaks.

- **Know where your household water comes from:** Is it from the city water supply? Most public water supplies are carefully monitored and treated to guard against contamination. Does a private well supply your water? Private water supplies are unlikely to be targets of intentional contamination. But they can become contaminated by accident and may not be as closely monitored as city water supplies.

- **Consider storing emergency water supplies in your home:** Store drinking water for circumstances in which the water supply may be polluted or disrupted. If water comes directly from a good, pretreated source,

25

then no additional purification is needed; otherwise, pretreat water before use. Store water in leak proof, break resistant containers. Consider using plastic bottles commonly used for juices and soft drinks. Keep water containers away from heat sources and direct sunlight.

- **Learn how to purify water:** Make sure that you include the supplies for this in your emergency kit: Knowing how to purify water is useful in any situation where you have to rely on untreated water.

Helpful tips to follow if there is an emergency affecting the water supply:

- **Follow all instructions from local authorities about purifying your water** (commonly called "boil orders") or using other water sources until authorities notify your community that it is safe to drink from the regular water supply again.

- **Do not strictly ration emergency drinking water supplies.** Try not to waste any water, but drink what you need. On average, a person needs about 2 qt (2 L) of water a day. Individual water needs vary depending on age, health, diet, and climate. Learn the signs of dehydration in children and adults so that you know what to watch for.

- **Use the safest water you have first** before turning to other water sources.

- If you know or suspect that your skin has come in direct contact with water that has been contaminated by a hazardous chemical or radiation fallout, follow the steps for personal decontamination to get the substance off your body as completely and quickly as possible.

The Decontamination Process - the following process can be used to decontaminate an individual:

- Remove the person from the contaminated area and into the decontamination corridor.

- Remove all clothing (at least down to their undergarments) and place the clothing in a labeled durable 6-mil polyethylene bag. Remove the contamination by rolling downward (from head to toe) and avoid pulling clothes off over the head.

- Thoroughly wash and rinse (using cold or warm water) the contaminated skin of the person using a soap and water solution. Be careful not to break the patient/victim's skin during the decontamination process, and cover all open wounds. Begin washing the person using soap and water solution and a soft brush. Always move in a downward motion (from head to toe). Make sure to get into all areas, especially folds in the clothing. Wash and rinse (using cold or warm water) until the contaminant is thoroughly removed.

- Cover the person to prevent shock and loss of body heat.

- Move the person to an area where emergency medical treatment can be provided.

HEALTHY LIVING

The Most Important Secrets You Must Learn In Order To Stay Healthy

Chapter 7: How to Purify Your Water

Water purification can greatly reduce your chance of getting sick from bacteria, viruses, and other living organisms in the water. You can disinfect water using one of the following methods:

- **Method # 1: Boiling water for one minute kills the microorganisms that cause disease. Directions:** Bring the water to a rolling boil for 1 minute. If you are at an elevation of 6500 ft (2000 m) or higher, boil the water for 3 minutes. This is the most effective purification method. However, it may be impractical if you need large quantities of water. It also requires a eat source, which you may not have in some emergencies. If fuel or power for your heat source is limited, bringing the water to a boil will usually disinfect it, even if you cannot boil it for the recommended time.

Remember: Do **NOT** Boil water to attempt to reduce nitrates. Boiling water contaminated with nitrates increases its concentration and potential risk. If you are concerned about nitrates, talk to your doctor about alternatives to boiling water for baby formula. You may choose to boil your water to remove microbial contaminants. Keep in mind that boiling reduces the volume of water by about 20 percent, thus concentrating those contaminants not affected by the temperature of boiling water, such as nitrates and pesticides.

- **Method # 2:** Add 16 drops of household liquid bleach for each gallon of water, stir, and let it stand for 30 minutes. If the water does not smell slightly like bleach after 30 minutes, add 16 more drops of bleach and let it stand for another 15 minutes. You should notice a bleach smell.

- **Method # 3:** Use iodine or chlorine purification tablets or drops. You can get these at stores that sell camping equipment and at some drugstores. Follow the instructions on the package. Purification tablets are not as effective as boiling or disinfecting with bleach. However, they do kill some types of organisms.

- **Method # 4: Activated carbon filters** adsorb **organic contaminants** that cause taste and odor problems. Depending on their design, some units can remove chlorination byproducts, some cleaning solvents, and pesticides. To maintain the effectiveness of these units, the carbon canisters must be replaced periodically. Activated carbon filters are efficient in removing metals such as lead and copper if they are designed to absorb or remove lead. There are many different types of filters, so be sure that you know what kinds of organisms your filter is designed to eliminate.

- **Method # 5: Ion exchange units** can be used to remove minerals from your water, particularly calcium and magnesium, they are sold for water softening. Some ion exchange softening units remove radium and barium from water. Ion exchange systems that employ activated alumina are used to remove fluoride and arsenate from water. These units must be regenerated periodically with salt.

- **Method # 6: Distilling water** - None of the purification methods described above eliminates heavy metals, salts, chemicals, or radioactive dust or dirt (fallout) from water. Many of these substances can be removed by distilling water, a more complicated method of purifying water.

- **Method # 7: Reverse osmosis treatment units** generally remove a more diverse list of contaminants than other systems. They can remove nitrates, sodium, other dissolved inorganics, and organic compounds.

Method # 8: Decrease Radioactive Results by Using a Homemade Filter

You will be surprised at how easy it is to make a homemade water filter. You will need a little time and you will need to be a little handy. However, for the average person it will be possible to construct a homemade water filter in more or less one hour.

One of the very big advantages of a homemade water filter is that you know what it is made of and you know you are giving your family the best possible protection.

How to make a homemade filter:

1. Punch holes in the bottom of a bucket, and cover the bottom with 1.5 in. (3.8 cm) of gravel. Cover the gravel with a towel. **Tip:** Make sure the sand and gravel have been washed before you put them in the filter. Otherwise, the sand and gravel will only make the water dirty.

2. Place the bucket over a larger container, and pour the water into the bucket so that it filters through the towel and gravel and drains into the container below.

3. Disinfect the water by boiling and adding chlorine bleach. You can filter the water by adding a couple drops of chlorine or by using purification tablets.

4. Replace the gravel after every 50 qt (47 L) of water.

HEALTHY LIVING

**The Most Important Secrets
You Must Learn
In Order To Stay Healthy**

Chapter 8: How Can You Benefit from Using a Whole House Water Filter?

A whole house water filter is often installed to make the water from any point-of-use in the house clean and safe. The owners of private houses more often address to whole house water filters, as they are independent to choose any kind of water purification system they want and to make changes to plumbing devices.

Those who live in apartment building have to coordinate the case with the local authorities as public water purification systems may already be installed in the building.

One way or another you can benefit a lot from installation of a whole house water filter. The most obvious reason to use this **method of water purification** is that clean, safe water flows from any tap in the house.

Chlorine that is necessary present in the water taken from the public water sources is eliminated just on its entering the plumbing system of the house. So chlorine can be released into the air. Being eliminated from the water for laundry purposes, chlorine cannot stick to the clothes. Otherwise, chlorine affects our body all the time through the clothes.

If chlorine is not removed from the water, it is released from the dishwasher into the air. The dishwasher is one of the main sources of chlorine vapors. Drinking water that was purified by means of a **whole house water filter** is of greater quality in general.

Whole house water filters are especially useful for those who suffer from asthma, breathing diseases and allergies. They need very clean air to prevent the development of their diseases.

However, not only does drinking water benefit from **whole house water filters**. Whole house water purification systems supply pure water to the shower. So you bath and shower in pure water.

Although separate shower water filters are available, a whole house water filter is more effective. Shower water filters have to work at a high temperature and not all the chlorine molecules are removed. The water purification process in whole house water filters is organized at a low temperature, which is a more efficient way to reduce chlorine. Due to these characteristics of whole house water filters, you are less likely to experience carcinogenic effects while drinking or inhaling chlorine and its byproducts.

HEALTHY LIVING

The Most Important Secrets
You Must Learn
In Order To Stay Healthy

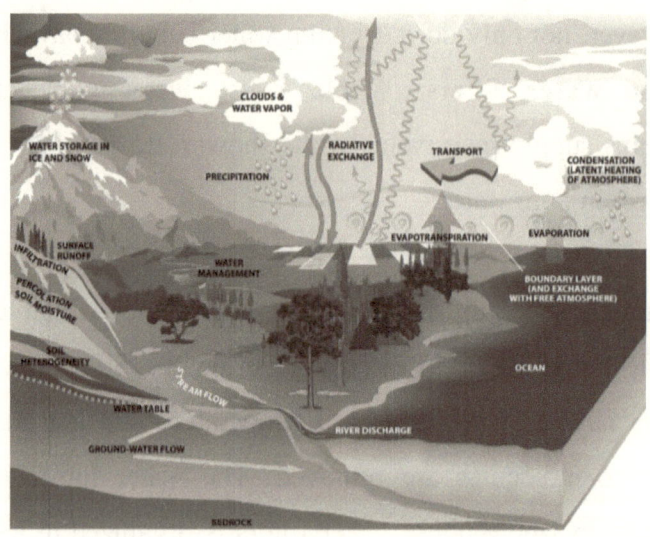

Chapter 9: How Safe Is the Drinking Water If You Get Your Water Supply from a Household Well?

EPA regulates public water systems; it does not have the authority to regulate private wells. Approximately 15 percent of Americans rely on their own private drinking water supplies (*Drinking Water from Household Wells,* 2002), and these supplies are not subject to EPA standards.

Unlike public drinking water systems serving many people, they do not have experts regularly checking the water's source and its quality before it is sent to the tap. These households must take special precautions to ensure the protection and maintenance of their drinking water supplies.

The Risk Involved

The risk of having problems depends on how good your well is—how well it was built and located, and how well you maintain it. It also depends on your local environment. That includes the quality of the aquifer from which your water is drawn and the human activities going on in your area that can affect your well. Several sources of pollution are easy to spot by sight, taste, or smell. However, many serious problems can be found only by testing your water. Knowing the possible threats in your area will help you decide the kind of tests you may need.

Here are six strategies you can take to help protect your private drinking water:

1. Identify potential problem sources.
2. Talk with local experts.

3. Have your water tested periodically.

4. Have the test results interpreted and explained clearly.

5. Set and follow a regular maintenance schedule for your well, and keep up-to-date records.

6. Immediately correct fix any problems

Track Any Problems Before They Become Big Problems

Understanding and spotting possible pollution sources is the first step to safeguarding your drinking water. If your drinking water comes from a well, you may also have a **septic system**. Septic systems and other on-site wastewater disposal systems are major potential sources of contamination of private water supplies if they are poorly maintained or located improperly, or if they are used for disposal of toxic chemicals. Information on septic systems is available from local health departments, state agencies, and the National Small Flows Clearinghouse *(www.epa.gov/owm/mab/smcomm/nsfc.htm)* at (800) 624-8301. A septic system design manual and guidance on system maintenance are available from EPA *(www.epa.gov/OW-WM.html/mtb/decent/homeowner.htm)*.

Have Your Water Tested Regularly

Test your water every year for total **coliform** bacteria, nitrates, total dissolved solids, and pH levels. If you suspect other contaminants, test for these as well. As the tests can be expensive, limit them to possible problems specific to your situation. Local experts can help you identify these contaminants. You should also test your water after replacing or repairing any part of the system, or if you notice any change in your water's look, taste, or smell. Often, county health departments perform tests for bacteria and nitrates. For other substances, health departments, environmental offices, or county governments should have a list of state-certified laboratories. Your State Laboratory Certification Officer can also provide you with this list. Call the Safe Drinking Water Hotline for the name and number of your state's certification officer. Any laboratory you use should be certified to do drinking water testing.

Important Factors to Remember In Order to Keep Your Water Purified and Your Family Safe

1. Periodically inspect exposed parts of the well for problems such as: Cracked, corroded, or damaged well casing, broken or missing well cap, and settling and cracking of surface seals.

2. Slope the area around the well to drain surface runoff away from the well.

3. Install a well cap or sanitary seal to prevent unauthorized use of, or entry into, the well.

4. Disinfect drinking water wells at least once per year with bleach or hypochlorite granules, according to the manufacturer's directions.

5. Have the well tested once a year for coliform bacteria, nitrates, and other constituents of concern.

6. Keep accurate records of any well maintenance, such as disinfection or sediment removal, that may require the use of chemicals in the well.

7. Hire a certified well driller for any new well construction, modification, or abandonment and closure.

8. Avoid mixing or using pesticides, fertilizers, herbicides, degreasers, fuels, and other pollutants near the well.

9. Do not dispose of wastes in dry wells or in abandoned wells.

10. Do not cut off the well casing below the land surface.

11. Pump and inspect septic systems as often as recommended by your local health department.

12. Never dispose of hazardous materials in a septic system.

After A Flood-Concerns and Advisories

1. Stay away from well pump to avoid electric shock.

2. Do not drink or wash from a flooded well.

3. Pump the well until water runs clear.

4. If water does not run clear, contact the county or state health department or extension service for advice.

HEALTHY LIVING

The Most Important Secrets You Must Learn In Order To Stay Healthy

Chapter 10: Learn How You Can Protect Your Water by Becoming an Advocate

Tips on how to protect the water we use:

- If you are one of the 15 percent of Americans who uses a private source of drinking water—such as a well, cistern, or spring—find out what activities are taking place in your **watershed** that may impact your drinking water; talk to local experts/ test your water periodically; and maintain your well properly.

- Find out if the Clean Water Act standards for your drinking water source are intended to protect water for drinking, in addition to fishing and swimming.

- Look around your watershed and look for announcements in the local media about activities that may pollute your drinking water.

- **Form and operate** a citizen's watch network within your community to communicate regularly with law enforcement, your public water supplier and wastewater operator. **Communication** is solution to a safer community!

- **Be alert**. Get to know your water/wastewater utilities, their vehicles, routines and their personnel.

- **Become aware of your surroundings -** This will help you to recognize suspicious activity as opposed to normal daily activities.

- **Attend public hearings** on new construction, storm water permitting, and town planning.

- **Keep your public officials accountable** by asking to see their environmental impact statements.

- **Ask questions** about any issue that may affect your water source.

- **Participate with your government** and your water system as they make funding decisions.

- **Volunteer** or help recruit volunteers to participate in your community's contaminant monitoring activities.

- **Help ensure** that local utilities that protect your water have adequate resources to do their job.

- **If you see any suspicious activities** in or around your water supply, please notify local authorities or call **9-1-1 immediately** to report the incident.

BE GREEN! Do not contaminate!

- **Reduce paved areas:** use permeable surfaces that allow rain to soak through, not run off.

- **Reduce or eliminate pesticide application:** test your soil before applying chemicals, and use plants that require little or no water, pesticides, or fertilizers.

- **Reduce the amount of trash you create:** reuse and recycle.

- **Recycle used oil:** 1 quart of oil can contaminate 2 million gallons of drinking water—take your used oil and antifreeze to a service station or recycling center.

- **Take the bus instead of your car one day a week:** you could prevent 33 pounds of carbon dioxide emissions each day.

- **Keep pollutants away from boat marinas and waterways:** keep boat motors well tuned to prevent leaks, select nontoxic cleaning products and use a drop cloth, and clean and maintain boats away from the water.

HEALTHY LIVING

The Most Important Secrets
You Must Learn
In Order To Stay Healthy

Glossary

Action Level - The level of lead and copper which, if exceeded, triggers treatment or other requirements that a water system must follow.

Aquifer - A natural underground layer, often of sand or gravel that contains water.

Bacteria – microorganisms that consist of a single chromosome and can cause severe house effects.

Botulism - a rare but very serious type of food poisoning caused by toxins produced by bacteria (*Clostridium botulinum*) that are commonly found in soil. Botulism is often caused by food that is not home-canned properly, such as home-canned beans and corn. In children younger than 1 year, botulism may be caused by bacteria found in raw (unpasteurized) honey or corn syrup. An adult's digestive system can defend against the bacteria in these foods, but an infant's digestive system cannot. Newborns and infants should not be given raw honey or corn syrup.

Symptoms of botulism usually begin 12 to 36 hours after the person eats contaminated food. Symptoms include blurred or double vision, muscle weakness, fatigue, dizziness, and headache. The person may also have nausea, vomiting, diarrhea, and abdominal pain. The most noticeable symptoms in children include double vision, irritability, and muscle weakness. Some children may have vomiting, constipation, inability to pass urine (urinary retention), and a dry mouth.

Botulism is potentially fatal and requires **immediate medical care**. People who have botulism will often be admitted to a hospital for treatment.

Cancer - a disease when bad cells grow uncontrolled in the body. Often is followed by death.

Carcinogens are substances that produce cancer.

Clean – the object is considered to be clean when it is free of rust, sediment, sludge and other contaminants.

Coliform - A group of related bacteria whose presence in drinking water may indicate contamination by disease-causing microorganisms

Community Water System (CWS) - A water system that supplies drinking water to 25 people or more year-round in their residences

Contaminant - Anything found in water (including microorganisms, radionuclides, chemicals, minerals, etc.) which may be harmful to human health

Cryptosporidium - Microorganism found commonly in lakes and rivers, which is highly resistant to disinfection.

Disinfectant - A chemical (commonly chlorine, chloramines, or ozone) or physical process (e.g., ultraviolet light) that kills microorganisms such as viruses, bacteria, and protozoa

Distribution System - A network of pipes leading from a treatment plant to customers' plumbing systems

E. Coli - (*Escherichia coli*) is a type of bacteria that normally lives in the digestive tract of humans and animals. Some kinds (strains) of *E. coli* can cause diarrhea and other digestive system (gastrointestinal) problems.

A few strains of *E. coli* bacteria (primarily a strain called O157:H7) can produce poisons (toxins) that can harm the intestines, blood, and kidneys. Bloody diarrhea commonly occurs when a person is infected with this strain of bacteria. A small number of infected people (especially small children and older adults) may develop serious, sometimes fatal, complications, usually from dehydration or severe blood and kidney problems.

Treatment of *E. coli* infection generally consists of managing complications, such as dehydration caused by diarrhea. Antidiarrheal medicines are not used because they may slow the elimination of the toxin from the body and increase the risk of complications. Antibiotics also may be harmful in cases of *E. coli* infection. The risk of illness from *E. coli* can be reduced by proper food preparation and storage and by careful hand-washing and other good personal hygiene habits.

Food Poisoning - an illness caused by eating foods that have harmful organisms in them. These harmful germs can include bacteria, parasites, and viruses. They are mostly found in raw meat, chicken, fish, and eggs, but they can spread to any type of food. They can also grow on food that is left out on counters or outdoors or is stored too long before you eat it. Sometimes food poisoning happens when people do not wash their hands before they touch food.

Most of the time, food poisoning is mild and goes away after a few days. All you can do is wait for your body to get rid of the germ that is causing the illness. However, some types of food poisoning may be more serious, and you may need to see a doctor.

Ground Water - Water pumped and treated from an aquifer

Hard Water is water that has too much positively charged ions of magnesium or calcium.

Hyperfiltration - another name for the process of reverse osmosis.

Inorganic Contaminants - Mineral-based compounds such as metals, nitrates, and asbestos; naturally occurring in some water, but can also enter water through human activities

Maximum Contaminant Level - The highest level of a contaminant that EPA allows in drinking water (legally enforceable standard)

Maximum Contaminant Level Goal - The level of a contaminant at which there would be no risk to human health (not a legally enforceable standard)

Microorganisms - Tiny living organisms that can be seen only under a microscope; some can cause acute health problems when consumed in drinking water

Organic Contaminants - Carbon-based chemicals, such as solvents and pesticides, which enter water through cropland runoff or discharge from factories

Particles are solids or liquids that are small and light enough to be suspended in the air.

Pure water – water that does not contain harmful substances and contaminants. Spring water is considered pure. Otherwise pure water can be get with the help of water purification systems.

Reverse Osmosis is a process sometimes referred to as hyperfiltration. Provides water of a very high quality. Reverse osmosis is based on the use of semi-permeable membrane that lets certain particles pass through it and traps others. Water moves through the membrane under pressure and all the contaminants larger in size than water molecules are removed. Pure water is collected on the other side of the membrane.

Septic System - Used to treat sanitary waste; can be a significant threat to water quality due to leaks or runoff

Shower water filter – is a water filter installed on a showerhead. Purifies the water that goes through the shower. Shower water filters especially effective for chlorine and its byproducts removal. Protect your skin and hair.

Toxic – harmful, hazardous.

Toxicant – poison, contaminant.

Tularemia - also called deerfly fever or rabbit fever, is a disease that usually occurs in animals. However, the disease can be passed to people through infected insects or animals or by exposure to contaminated water or dust.

Humans are most commonly infected by the following:

- Being bitten by a tick
- Being bitten by a deerfly
- Being bitten by a mosquito
- Skinning, dressing, or handling diseased animals.
- Drinking water that is contaminated with urine or feces
- Inhaling contaminated dust

This disease is found throughout the United States, but most cases are reported in Arkansas, Missouri, and Oklahoma. Symptoms usually start within 21 days (but average 1 to 10 days) after the tick bite or other exposure. Symptoms of tularemia include:

- Chills and high fever up to 106F (41.1C), often starting suddenly.
- Headache that is often severe.
- An open craterlike sore (ulcer) at the site of the bite
- Swollen glands near the site of the bite

- Nausea and vomiting

Prescription medicine is used to treat tularemia.

Vapors – we speak about vapors in connection with the gaseous phase of substances, especially of those that are liquid or solid in normal conditions: at normal atmospheric pressure and temperature.

Viruses are tiny life forms. They spread a large variety of human infections.

Vulnerability Assessment - An evaluation of drinking water source quality and its vulnerability to contamination by pathogens and toxic chemicals

Water purification – the process during which contaminants are removed from water. Water purification can take place either in public water sources or in the houses with the help of water filters.

Well - A bored, drilled or driven shaft whose depth is greater than the largest surface dimension, a dug hole whose depth is greater than the largest surface dimension, an improved sinkhole, or a subsurface fluid distribution system

Watershed - The land area from which water drains into a stream, river, or reservoir

Whole house water filter is a water filter that is installed in the place where water enters your house. Whole house water filter makes the water from all the taps in the house safe to drink, to bathe in and to use for domestic purposes.

Resources

American Water Works Association
Public Affairs Department
6666 West Quincy Avenue
Denver, CO 80235
Phone (303) 794-7711
www.awwa.org

Association of Metropolitan Water Agencies
1620 I Street NW
Suite 500
Washington, DC 20006
Phone (202) 331-2820
Fax (202) 785-1845
www.amwa.net

Association of State Drinking Water Administrators
1401 Wilson Blvd.
Suite 1225
Arlington, VA 22209
Phone (703) 812-9505
www.asdwa.org

Clean Water Action
4455 Connecticut Avenue NW Suite A300
Washington, DC 20008
Phone (202) 895-0420
www.cleanwater.org

Consumer Federation of America
1620 I Street NW

Suite 200
Washington, DC 20006
Phone (202) 387-6121
www.consumerfed.org

The Groundwater Foundation
P.O. Box 22558
Lincoln, NE 68542
Phone (800) 858-4844
www.groundwater.org

The Ground Water Protection Council
13308 N. Mac Arthur
Oklahoma City, OK 73142
Phone (405) 516-4972
www.gwpc.org

International Bottled Water Association
1700 Diagonal Road
Suite 650
Alexandria, VA 22314
Phone (703) 683-5213
Information Hotline 1-800-WATER-11
ibwainfo@bottledwater.org

National Association of Regulatory Utility Commissioners
1101 Vermont Ave NW
Suite 200
Washington, DC 20005
Phone (202) 898-2200
www.naruc.org

National Association of Water Companies
2001 L Street NW
Suite 850

Washington, DC 20036
Phone (202) 833-8383
www.nawc.org

National Drinking Water Clearinghouse
West Virginia University
P.O. Box 6064
Morgantown, WV 26506
Phone (800) 624-8301
www.ndwc.wvu.edu

National Ground Water Association
601 Dempsey Rd
Westerville, OH 43081-8978
Phone: (800) 551-7379
www.ngwa.org

National Rural Water Association
2915 South 13th Street
Duncan, OK 73533
Phone (580) 252-0629
www.nrwa.org

Natural Resources Defense Council
40 West 20th Street
New York, NY 10011
Phone (212) 727-2700
www.nrdc.org

NSF International
P.O. Box 130140
789 North Dixboro Road
Ann Arbor, MI 48113
Phone (800) NSF-MARK
www.nsf.org

Rural Community Assistance Program
1522 K Street NW
Suite 400
Washington, DC 20005
Phone (800) 321-7227
www.rcap.org

Underwriters Laboratories
Corporate Headquarters
2600 N.W. Lake Road
Camas, WA 98607
Phone (877) 854-3577
www.ul.com

Water Quality Association
4151 Naperville Road
Lisle, IL 60532
Phone (630) 505-0160
www.wqa.org

U.S. Environmental Protection Agency Water
Resource Center
1200 Pennsylvania Avenue NW
RC-4100T
Washington, DC 20460
SDWA Hotline (800) 426-4791
www.epa.gov/safewater

Water Systems Council
National Programs Office
101 30th Street NW
Suite 500
Washington, D.C. 20007
Phone: (202) 625-4387
Wellcare Hotline 888-395-1033
www.watersystems council.org

About the Author

Stacey Chillemi

Stacey Chillemi graduated from Richard Stockton College in Pomona, New Jersey, majoring in marketing and advertisement. In the mid-nineties while in college, she began her first book, *Epilepsy: You're Not Alone*. It was published six years later. Before and after graduation in 1996, she worked in New York City for NBC. Since the birth of her children, she has been a freelance journalist.

She has written features for journals and newspapers. Her articles have appeared in dozens of newspapers and magazines in North America and abroad. She won an award from the Epilepsy Foundation of America in 2002 for her help and dedication to people with epilepsy.

My Web Sites: http://www.lulu.com/spotlight/staceychil
http://www.authorsden.com/staceydchillemi

BOOKS PUBLISHED BY STACEY CHILLEMI:

1. The Complete Herbal Guide: A Natural Approach to Healing the Body

2. How to Live Comfortably with Asthma
3. Epilepsy You're Not Alone
4. Eternal Love: Romantic Poetry Straight from the Heart
5. My Mommy Has Epilepsy (Children's Book)
6. My Daddy Has Epilepsy (Children's Book)
7. Keep the Faith: To Live and Be Heard from the Heavens Above (poetry book)
8. Live, Learn, and Be Happy with Epilepsy
9. Epilepsy and Pregnancy: What Every Woman Should Know
   Co-authored by Dr. Blanca Vasques
10. Faith, Courage, Wisdom, Strength and Hope
11. How to Be Wealthy Selling Informational Products on the Internet
12. How to Become Wealthy in Real Estate
13. How to Become Wealthy Selling Ebooks
14. Life's Missing Instruction Manual: Beyond Words
15. How To Become Wealthy Selling Products on The Internet
16. Breast Cancer: Questions, Answers & Self-Help Techniques
17. How Thinking Positive Can Make You Successful: Master The Power Of Positive Thinking

STACEY CHILLEMI STORIES AND POETRY HAVE BEEN PUBLISHED IN:

- Chicken Soup for the Recovering Soul
- Chicken Soup for the Shoppers Soul
- Whispers of Inspiration

ACCOMPLISHMENTS:

- Book Signing at Borders in Freehold, New Jersey for Faith, Courage, Wisdom, Strength and Hope" – July 2009
- Writer for Neurology Now Magazine (The Academic Academy of Neurology – The Epilepsy Column) February 2010
- February 2010, Wrote an article about Epilepsy & Menstruation with Dr. Devinsky-Epileptologist from NYU)
- H.O.P.E. Mentor, for the Epilepsy Foundation
- Speaker at different events for schools, organizations, political events
- Spoke in front of Congress in Washington for employment discrimination for people with epilepsy
- Appeared on four talk shows to discuss epilepsy focusing on the importance of understanding epilepsy, how to help someone having a seizure and giving people with epilepsy encouragement and hope for the future.
- Appeared on radio stations discussing epilepsy
- Appeared on the Michael Dressor Show – Health Radio
- Appeared in newspapers all over New Jersey such as, The Leader, Belleville Post and the Star Ledger.
- June 26, 2002, I was honored an award by the Epilepsy Foundation of New Jersey for Outstanding Volunteer Award.
- Received awards in my achievements and certificates in recognition for outstanding efforts in trying to improve society.
- Active participant in organizations and activities.
- Published over 400 articles.
- Author has a dynamic personality and strong public speaking skills.

EDUCATION:

Stockton College, Pomona NJ
- Speaker at different events for schools, organizations, political events
- Associate Degree in Business
- BA in Marketing
- Minor in Advertising

CAREER EXPERIENCE:

- Journalist for The Journal Magazine
- Worked for NBC on Dateline
- Channel 4 News
- Today Show
- Managing Editor for the Fashion Magazine **UZURI**.

References

1. 2011 Healthwise, Incorporated. Healthwise, Healthwise for every health decision, E. Gregory Thompson, MD - Internal Medicine, Christine Hahn, MD - Epidemiology
2. U.S. Food and Drug Administration (2007). FDA approves first U.S. vaccine for humans against the avian influenza virus H5N1. FDA News. Available online: http://www.fda.gov/NewsEvents/Newsroom/PressAnnouncements/2007/ucm108892.htm.
3. American Red Cross (2010). Terrorism. Available online: http://www.redcross.org/portal/site/en/menuitem.86f46a12f382290517a8f210b80f78a0/?vgnextoid=cbc95d795323b110VgnVCM10000089f0870aRCRD&vgnextfmt=default.
4. Centers for Disease Control and Prevention (2010). Emergency Preparedness and Response. Available online: http://www.bt.cdc.gov.
5. U.S. Department of Homeland Security (2010). Be Informed. Available online: http://www.ready.gov/america/beinformed/index.html.
6. W. David Colby IV, MSc, MD, FRCPC - Infectious Disease
7. Centers for Disease Control and Prevention (2010). Campylobacter. Available online: http://www.cdc.gov/nczved/divisions/dfbmd/diseases/campylobacter.
8. Anderson WT (2004). Food-borne and water-borne diseases. In JE Tintinalli et al., eds., Emergency Medicine: A Comprehensive Study Guide, 6th ed., pp. 964-969. New York: McGraw-Hill.

9. Center for Food Safety and Applied Nutrition, U.S. Food and Drug Administration (2009). Foodborne Pathogenic Microorganisms and Natural Toxins Handbook. Available online: http://www.fda.gov/Food/FoodSafety/FoodborneIllness/FoodborneIllnessFoodbornePathogensNaturalToxins/BadBugBook/default.htm.
10. Centers for Disease Control and Prevention (2004). Diagnosis and management of foodborne illness. A primer for physicians and other health care professionals. MMWR, 53(RR-4): 1-32.
11. Centers for Disease Control and Prevention (2008). Toxoplasmosis. Available online: http://www.cdc.gov/toxoplasmosis/factsheet.html.
12. Centers for Disease Control and Prevention (2009). Foodborne infections. Available online: http://www.cdc.gov/nczved/divisions/dfbmd/diseases/foodborne_infections.
13. Centers for Disease Control and Prevention (2009). Listeriosis. Available online: http://www.cdc.gov/nczved/divisions/dfbmd/diseases/listeriosis.
14. Centers for Disease Control and Prevention (2009). Noroviruses and drinking water from private wells. Available online: http://www.cdc.gov/healthywater/drinking/private/wells/disease/norovirus.html.
15. Centers for Disease Control and Prevention (2009). Salmonellosis. Available online: http://www.cdc.gov/nczved/divisions/dfbmd/diseases/salmonellosis.
16. Centers for Disease Control and Prevention (2009). Shigellosis. Available online: http://www.cdc.gov/nczved/divisions/dfbmd/diseases/shigellosis.

17. Centers for Disease Control and Prevention (2010). Campylobacter. Available online: http://www.cdc.gov/nczved/divisions/dfbmd/diseases/campylobacter.
18. Centers for Disease Control and Prevention (2010). Cryptosporidiosis (also known as "crypto"). Available online: http://www.cdc.gov/crypto.
19. Centers for Disease Control and Prevention (2010). <u>Escherichia coli</u>. Available online: http://www.cdc.gov/nczved/divisions/dfbmd/diseases/ecoli_o157h7/index.html.
20. Centers for Disease Control and Prevention (2010). Marine toxins. Available online: http://www.cdc.gov/nczved/divisions/dfbmd/diseases/marine_toxins.
21. Food Safety and Inspection Service (2006). Foodborne illness: What consumers need to know. Available online: http://www.fsis.usda.gov/fact_sheets/Foodborne_Illness_What_Consumers_Need_to_Know/index.asp.
22. Sodha SV, et al. (2010). Foodborne disease. In GL Mandell et al., eds., <u>Mandell, Douglas, and Bennett?s Principles and Practice of Infectious Diseases,</u> 7th ed., vol. 1, pp. 1413-1427. Philadelphia: Churchill Livingstone Elsevier.
23. U.S. Department of Agriculture Food Safety and Inspection Service (2006). Fact sheet. Safe food handling: Basics for handling food safely. Available online: http://www.fsis.usda.gov/fact_sheets/Basics_for_Handling_Food_Safely/index.asp.
24. (PDF) <u>*Combating Waterborne Diseases at the Household Level*</u>. <u>World Health Organization</u>. 2007. Part 1. ISBN <u>978-92-4-159522-3</u>. http://www.who.int/water_sanitation_health/publications/combating_disease part1lowres.pdf.

25. (PDF) *Water for Life: Making it Happen*. World Health Organization and UNICEF. 2005. ISBN 92-4-156293-5. http://www.who.int/water_sanitation_health/waterforlife.pdf.
26. Chen, Jimmy, and Regli, Stig. (2002). "Disinfection Practices and Pathogen Inactivation in ICR Surface Water Plants." *Information Collection Rule Data Analysis.* Denver:American Water Works Association. McGuire, Michael J., McLain, Jennifer L. and Obolensky, Alexa, eds. pp. 376-378. ISBN 1-58321-273-6
27. Aeration and Gas Stripping, Accessed June 4, 2012.
28. CO2 Degasifiers/Drinking Water Corrosion Control, Accessed June 4, 2012.
29. Degassing Towers, Accessed June 4, 2012.
30. American Water Works Association RTW corrosivity index calculator, Accessed June 4, 2012.
31. Edzwald, James K., ed. (2011). *Water Quality and Treatment.* 6th Edition. New York:McGraw-Hill. ISBN 978-0-07-163011-5
32. [a] [b] [c] [d] Crittenden, John C., et. al., eds. (2005). *Water Treatment: Principles and Design.* 2nd Edition. Hoboken, NJ:Wiley. ISBN 0-471-11018-3
33. [a] [b] Kawamura, Susumu. (2000). *Integrated Design and Operation of Water Treatment Facilities.* 2nd Edition. New York:Wiley. pp. 74-5, 104. ISBN 0-471-35093-1
34. United States Environmental Protection Agency (EPA)(1990). Cincinnati, OH. "Technologies for Upgrading Existing or Designing New Drinking Water Treatment Facilities." Document no. EPA/625/4-89/023.
35. [a] [b] Andrei A. Zagorodni (2007). *Ion exchange materials: properties and applications*. Elsevier. ISBN 978-0-08-044552-6.

http://books.google.com/books?id=XfDFKl3I74MC. Retrieved 22 November 2011.
36. Neemann, Jeff; Hulsey, Robert; Rexing, David; Wert, Eric (2004). "Controlling Bromate Formation During Ozonation with Chlorine and Ammonia." *Journal American Water Works Association.* 96:2 (February) 26-29.
37. Conroy RM, Meegan ME, Joyce T, McGuigan K, Barnes J (1999 October). "Solar disinfection of water reduces diarrhoeal disease, an update". *Arch Dis Child* **81** (4): 337–8. doi:10.1136/adc.81.4.337. PMC 1718112. PMID 10490440. //www.ncbi.nlm.nih.gov/pmc/articles/PMC1718112/.
38. Conroy RM, Meegan ME, Joyce TM, McGuigan KG, Barnes J (2001). "Use of solar disinfection protects children under 6 years from cholera". *Arch Dis Child* **85** (4): 293–5. doi:10.1136/adc.85.4.293. PMC 1718943. PMID 11567937. //www.ncbi.nlm.nih.gov/pmc/articles/PMC1718943/.
39. Rose A. at al. (2006 February). "Solar disinfection of water for diarrhoeal prevention in southern India". *Arch Dis Child* **91** (2): 139–41. doi:10.1136/adc.2005.077867. PMC 2082686. PMID 16403847. //www.ncbi.nlm.nih.gov/pmc/articles/PMC2082686/.
40. Hobbins M. (2003). The SODIS Health Impact Study, Ph.D. Thesis, Swiss Tropical Institute Basel
41. B. Dawney and J.M. Pearce "Optimizing Solar Water Disinfection (SODIS) Method by Decreasing Turbidity with NaCl", *The Journal of Water, Sanitation, and Hygiene for Development* 2(2) pp. 87-94 (2012). open access

42. Sciacca F, Rengifo-Herrera JA, Wéthé J, Pulgarin C (2010-01-08). "Dramatic enhancement of solar disinfection (SODIS) of wild Salmonella sp. in PET bottles by H(2)O(2) addition on natural water of Burkina Faso containing dissolved iron" (epub ahead of print). *Chemosphere* **78** (9): 1186–91. doi:10.1016/j.chemosphere.2009.12.001. PMID 20060566.
43. Centers for Disease Control and Prevention (2001). "Recommendations for using fluoride to prevent and control dental decay caries in the United States". *MMWR Recomm Rep* **50** (RR-14): 1–42. PMID 11521913. http://cdc.gov/mmwr/preview/mmwrhtml/rr5014a1.htm. Lay summary – *CDC* (2007-08-09).
44. Division of Oral Health, National Center for Prevention Services, CDC (1993) (PDF). *Fluoridation census 1992*. http://cdc.gov/fluoridation/pdf/statistics/1992.pdf. Retrieved 2008-12-29.
45. Reeves TG (1986). "Water fluoridation: a manual for engineers and technicians" (PDF). Centers for Disease Control. http://www.cdph.ca.gov/certlic/drinkingwater/Documents/Fluoridation/CDC-FluoridationManual-1986.pdf. Retrieved 2008-12-10.
46. US EPA emergency disinfection recommendations. Epa.gov. Retrieved on 2011-11-22.
47. Savage, Nora; Mamadou S. Diallo (1 October 2005). "Nanomaterials and Water Purification: Opportunities and Challenges". *J. Nanopart. Res.* **7** (4–5): 331–342. doi:10.1007/s11051-005-7523-5. http://www.ph.ucla.edu/ehs/ehs280/articles/Savage%20Diallo%20Review%20Nanotechnology%20Water%20Quality%20%28Hoek%202%29.pdf. Retrieved 24 May 2011.

48. Hydrates for Gypsum Stack Water Purification
49. " Miranda, Kim, Hull, et.a. "Changes in Blood Lead Levels Associated with Use of Chloramines in Water Treatment Systems" 03/13/2007. Environmental Health Perspectives.
50. Health risks from drinking demineralised water. (PDF) . Rolling revision of the WHO Guidelines for drinking-water quality. World Health Organization, Geneva, 2004
51. Kozisek F. (2004). Health risks from drinking demineralised water. WHO.
52. Water Distillers – Water Distillation – Myths, Facts, etc. Naturalsolutions1.com. Retrieved on 2011-02-18.
53. Minerals in Drinking Water. Aquatechnology.net. Retrieved on 2011-02-18.
54. 1 Factoids: Drinking Water & Ground Water Statistics for
55. 2002, 2003.
56. 2 Community Water Systems Survey 2000, Volume I, 2001.
57. 3 The Clean Water and Drinking Water Infrastructure Gap
58. Analysis, EPA 816-R-02-020.
59. 4 Factoids: Drinking Water and Ground Water Statistics for
60. 2001, EPA 816-K-02-004.
61. 5 Morbidity and Mortality Weekly Report: Surveillance for
62. Waterborne Disease Outbreaks, United States 1999-2000,
63. 2002.
64. 25 Years of the Safe Drinking Water Act, 1999
65. The Emergency Response Safety and Health Database (CDC/NIOSH)

ISBN: 978-1-300-24958-0

www.ingramcontent.com/pod-product-compliance
Lightning Source LLC
Chambersburg PA
CBHW021908170526
45157CB00005B/2014